CLASSIC AMERI LIMOUSINES
1955 THROUGH 2000
PHOTO ARCHIVE

Richard J. Conjalka

Iconografix
Photo Archive Series

Iconografix
PO Box 446
Hudson, Wisconsin 54016 USA

Library of Congress Card Number: 00-135946

ISBN 1-58388-041-0

01 02 03 04 05 06 07 5 4 3 2 1

Printed in the United States of America

Cover and book design by Shawn Glidden

Copy editing by Dylan Frautschi

COVER CAPTION: In 1989, Lincoln again became the official "Car of Presidents" with this Lincoln Town Car "Presidential Edition." Using a 460-cubic-inch engine from Ford's truck line, the 22-foot long limousine, which had a 3-year build time, was a reality. Wheelbase was a whopping 162 inches and the overall length was 263 inches. O'Gara-Hess & Eisenhardt in conjunction with Ford Motor Company were the builders of this limousine. Reportedly, the cost exceeded over $600,000.00. The rear compartment carried six adults in comfort and was first used on inauguration day in January 1989, for President George Bush. This limousine today is still used as a "back-up" vehicle for the President and various foreign visiting dignitaries.

Book Proposals

Iconografix is a publishing company specializing in books for transportation enthusiasts. We publish in a number of different areas, including Automobiles, Auto Racing, Buses, Construction Equipment, Emergency Equipment, Farming Equipment, Railroads & Trucks. The Iconografix imprint is constantly growing and expanding into new subject areas.

Authors, editors, and knowledgeable enthusiasts in the field of transportation history are invited to contact the Editorial Department at Iconografix, Inc., PO Box 446, Hudson, WI 54016.

Table of Contents

DEDICATION

This book is dedicated to the memory of my father, Joseph C. Conjalka, who knew I had a fascination with these automobiles ever since I was a child. At the time of his passing, he inspired me to author this work.

ACKNOWLEDGMENTS

The photographs that appear in this book came largely from the personal collection of car buff, Richard J. Conjalka. A significant number of these photos were acquired over the past fifty years from many automobile manufacturers and specialized coachbuilders, whom either exist today or have long closed their doors. Many thanks are in order to the following other people who have assisted in compiling the following photographs. They are as follows:

Mr. Mike McKiernan, Public Relations Department, Cadillac Motor Car Division of GM.
Tania Kordis, V.P. Marketing, Federal Coach, Fort Smith, Arkansas.
Mr. Gregg D. Merksamer, automotive Journalist and Author.
Mr. Jan Witte, Chrysler Imperial Car Club of America.
Also, to a select few members of the Professional Car Society enthusiasts who keep and preserve these classic automobiles for all to admire now and in the future.

Very rare and exquisitely preserved, this 1915 Brewster Limousine found a new home for just $55,000.00 at the inaugural New York Auto Salon and Auction at the Waldorf Astoria Hotel in New York City, in October of 1999. When this limousine was brand new in 1915, its price was only several thousand dollars. This car was one of the first motorized formal limousines, making the transformation from horse carriage transportation to motorized limousine transportation.

INTRODUCTION

American limousines have long been a part of our nation's culture–ever since the first motorized vehicle was in its infancy well over a hundred years ago.

Prior to the days of the gasoline or electric powered vehicles, horse-drawn carriages transported the elite and the wealthy. Known as Broughams or Laundaulets, these carriages were the forerunners of today's modern limousine that we see in every major metropolitan city in the United States.

Limousines are used today everywhere from the White House in our nation's capitol, on down to the "common person" who can easily afford to rent or lease one for a formal or special occasion, or just for "a night out on the town."

Ever since the day of the first motorized cars, limousines have been used for a myriad of reasons and purposes. For example, for transporting the wealthy and the elite of our society, for executive and corporate use because driving time for executives becomes "down time" for corporate productivity, for funeral home use in transporting bereaved families, for hotel courtesy shuttles and for anyone who wishes to use a formal vehicle for hire. Special occasions for a limousine-for-hire may include weddings, proms, funerals, or even a "Hollywood gala," such as the Oscars on Oscar night, which is held annually in Hollywood, California.

A limousine is generally defined as a vehicle which is large in size, has an enclosed rear compartment that can easily carry seven to nine passengers in comfort and has an open or closed roof over the driver's seat.

In the past when limousines were classic formal four-door cars, they were viewed with awe as cars of the rich and had a very impressive stature. They were often used by funeral homes and bankers or others of wealth, power and prominence.

The familiar Cadillac Fleetwood "75" Series nine-passenger sedan and limousine accounted for well over eighty-five percent of all limousines built during the post war period of the late 1940s and 1950s. Packard, who fought with Cadillac in this time period for sales, eventually died with a whimper. And so by 1955, Cadillac was the major manufacturer of all limousines in the U.S. Cadillac, being the dominant player in the field since WWII, remained the only assembly line production long wheelbase limousine manufacturer through the mid-eighties.

Of the "big three" American luxury cars left in this time period, Lincoln and Chrysler Imperial also tried to make a niche in the market. As hard as they tried, their sales of limousine units would never come close to Cadillac's. All Lincoln and Chrysler Imperial limousines were converted by specialized coachbuilders to special order only. Therefore, their production runs were rather limited.

Although Lehman-Peterson built the first stretch limousines in 1963, it was not until the late 1970s that the stretch "limo" became a fairly common phenomenon. At this time many specialty coachbuilders began turning out custom-built stretch limousines. Some of these stretch jobs began to find their way into daily use. In 1977 many limousine operators realized that the new factory downsized Cadillac limousine was just too small and had disadvantages due to its size. Ease of entry and egress was difficult because of the short wheelbase and jump seats. The car was no longer practical for use in the corporate, funeral and livery services. Thus, was born the stretch.

At one time in the 1980s, there were almost 40 specialized coachbuilders. They would literally cut in half a four-door or two-door car to create a custom limousine suited to end users specifications and delight. "The sky was the limit" as to equipment and custom touches when ordered.

The limousine as we know it today has evolved to almost shuttle bus size proportions carrying anywhere from twelve to fifteen people. These are known as party or bar cars. They are a far cry from the formal classics of the past, which had a short wheelbase and folding jump seats. The formal limousine, as many "old timers" remember it, was popular in the last three quarters of the twentieth century. Now they are almost extinct, as only a few are produced annually, for the United States Government and the Secret Service Branch. These "special limousines" are built to very strict Secret Service specifications and standards and are not publicized in any way. Specially built by several coachbuilders, who are experts in body armoring, these cars are rolling fortresses on wheels.

Many custom and classic limousines are still in existence today, thanks to collectors, auto museums, funeral homes and others who appreciate these cars. The Professional Car Society, whose members are dedicated to collecting, restoring and showing these types of cars are one such group of people. PCS membership is worldwide and many of these classic and formal limousines show up at meets around the country annually.

The Professional Car Society's address is:
P.O. Box 9636
Columbus, Ohio 43209

The "standard of the world" for 1955, Cadillac Fleetwood "75" sedans and limousines were known as "the Flagships of the Fleet." Prices depended on whether an eight-passenger sedan or a limousine with glass divider was ordered. They were respectively $5,694.00 or $5,896.00. Approximately 1,850 sedans and limousines were produced in 1955. The car shown here is a good example of early to mid-fifties limousines and Cadillacs, which outsold all other limousines manufactured in the 1950s.

The 1956 Cadillac limousine was the last of the "Fishtail fin" styling series. These fins had been used on Cadillacs since 1948. Since this was America's only true production limousine in the 1950s, wherever people of prominence or wealth gathered you were sure to see a chauffeur-driven Fleetwood "75" sedan or limousine. The year 1956 boasted over 1,100 of these examples. Prices were about $6,300.00 (air-conditioning cost extra).

The 1957 Cadillac line-up, including the "75" Series, was completely restyled. The rear roofline still followed the same styling introduced in 1950. Since Cadillac introduced its ultra luxurious El Dorado Brougham show car in this year the "75" was no longer the "flagship" car. Combined sedan and limousine production totaled 1,900 and prices were in the $7,500.00 range, while the handmade El Dorado Brougham's price was $13,084.00. Production for the El Dorado Brougham totaled only 400 for Cadillac from 1957 to 1960.

For the Cadillac "75" Series, 1958 would be the last year for the antiquated styling introduced in 1955. Slightly over 1,500 cars were produced in 1958. Prices ranged from $8,310.00 to $8,525.00. The entire 1958 line, including the "75s," shared common styling. The 1959 line would see brand new crisp styling.

The 1959 Cadillac "75" Series had completely new styling cues from bumper to bumper. It was the first major restyling since the early 1950s. The awe-inspiring '59s had the highest tail fin peak of any car produced in the 1950s. The new styling was crisp and angular for this new generation of Fleetwood "75s," and would continue through the 1965 models.

The imposing "75" Series Cadillac Fleetwood got a fresh and slightly more restrained styling for 1960. More sedate horizontal lines and a crisp roofline contributed to the formal appearance of this car. A little over 1,500 units were produced. Base price was $9,750.00 and rose from there. Air-conditioning was extra–can you believe that?

At the top of the 12-model line-up for 1961 was this stately Fleetwood Limousine. Styling in 1961 was again all new and redone from the previous year. These large cars still rode on the same 149.8-inch wheelbase as in previous years. This limousine cost $9,745.00 and only 926 were built, while 699 of the sedan version, without the glass divider, were produced.

The 1962 Cadillac Limousine was almost identical in appearance to the 1961 model. Both the nine-passenger sedan and the limousine retained the same upper structure and windshield as used on all 1959-1965 models. In the limousine version the chauffeur's compartment was upholstered in black or gray leather while the rear fabric was a gray material called broadcloth.

The 1965 Cadillac Fleetwood "75" Series nine-passenger sedan and limousine were almost identical to the 1964 model except for the wheel covers. This was the last year of this venerated body style as the 1966 would be all new. The 1965 model "75" Series styling was not updated, even though the rest of the Cadillac line-up was. The 1965 Fleetwood "75" retailed for almost $10,000.00 and weighed 5,260 pounds. Production was about 1,200 units. A landau roof, which blocked out the rear quarter windows, was an extra cost option.

In 1966, the prestigious Cadillac Fleetwood "75" sedan and limousine underwent their first total restyling since 1959. Built on a 149.8-inch wheelbase, this was an unusually handsome car that retailed for $10,521.00. Combined sales, for both the sedan version and the limousine, were close to 2,000.

In the year of 1967, Cadillac offered 12 cars in its model line-up. The 1967 Fleetwood Limousine was an impressive car with folding jump seats and doors that opened into the roof for ease of entry and egress. The cost was equally impressive for that year. Limousines were now $10,360.00 and only 835 were produced as nine-passenger sedans. The author can remember one that was used for carrying bereaved families at the funeral home where he worked that year–black in color with gray interior broadcloth. In the rear, Kleenex was provided for tears.

18

The 1968 Cadillac Fleetwood "75" limousines and nine-passenger sedans were almost identical to the 1967 models. Carrying nine adults in full comfort, this car rode on a full 149.8-inch wheelbase and was powered by a huge 472-cubic-inch engine. This car was widely used in many funeral homes for bereaved families.

The author photographed these 1974 Cadillac Fleetwood "75" limousines, in February 1974, at the annual Chicago auto show held at McCormick Place Exhibition Hall. These limousines rode on a long 149.8-inch wheelbase and were powered by a huge 472-cubic-inch engine. Premium gasoline was definitely to be used–even at this time of the Middle East oil scare. Production totaled close to 1,900 units and retailed at about $12,500.00. For $2,500.00 more, a formal landau roof with Landau Irons could be ordered, bringing the cost of the limousine to over $14,000.00.

A 1975 Cadillac Seville Limousine, manufactured by Universal Coach of Michigan, is photographed here in the snow. Universal Coach produced specialty cars to order only. Very few of these were ever produced in mass quantities. This is a "double-cut" stretch Seville. The center section plus the extended quarter panel behind the rear door gave the occupants the utmost rear room in a Seville Cadillac. Note the blue matching vinyl roof.

The last of the big "full size" factory Cadillac formal cars was this 1976 Cadillac Fleetwood "75" nine-passenger sedan and limousine. 1975 and 1976 were the only two years in which the 151.5-inch wheelbase was powered by a huge 500-cubic-inch engine. This car retailed for $15,240.00 in limousine form. Production totaled around 1,800 units for this year. This is the "last generation" of the "big ones" from the factory in Detroit. Note the chrome wire wheel cover.

The year 1977 brought the first downsized Cadillac as GM began downsizing their entire line of cars. This was Phaeton Coaches' (of Dallas, Texas) answer to the downsized Cadillac Limousine. Here's their downsized 26-inch stretch Cadillac limousine built on a new Fleetwood Brougham chassis. Notice there is no partition divider. It was powered by Cadillac's new 425-cubic-inch motor. This was one of the very first limousines to be stretched on the new downsized chassis.

New York Custom Coach, in 1982, produced this unusual Fleetwood Cadillac Limousine stretch conversion, with the extension in the rear doors and quarter panel. Seating in the rear interior was face to face with a center console housing beverage and entertainment consoles. This car has a total 42-inch stretch including the doors. Notice the chrome roof, opera light, wreath and crest on the upper roof, along with the "blind spot rear quarters."

1983 was a good year for Superior Coach of Lima, Ohio. A major player in the funeral coach and limousine market, Superior produced this 1983 Cadillac six-door limousine. The conversion added the center door and allowed 3 rows of seats. Six-doors were primarily built for funeral homes serving the families of the deceased. They were much easier to enter or egress than the traditional four-door car.

Last of the factory rear-wheel drive formal cars. This 1984 Cadillac Fleetwood "75" nine-passenger sedan and limousine was the last of the full size production line of in-house limousines built by Cadillac. This car rode on a 144.5-inch wheelbase. Prices exceeded $30,000.00 for this last factory formal, now coveted by collectors.

Cadillac boasted many coachbuilders in 1984. This particular model was the Moloney Coachbuilders' "grand flagship" limousine. It is seen here at a church wedding in Indiana. Notice the classic "continental spare" fifth wheel. This limousine was a double-cut stretch and was equipped with every conceivable creature comfort.

The 1984 Manhatten Formal. From 1928 to 1984, Armbruster-Stageway, Inc. did a number of these interesting "Manhatten Formal Limousines" for those that wanted a Fleetwood "75" Cadillac plus 12 inches more room in the center for additional jump seat passenger leg room. Located in Fort Smith, Arkansas, Armbruster-Stageway, Inc. built quite a few different conversions, on all types of makes and models, into limousines, airporters, and funeral cars.

The "New Generation" Cadillac Limousine was custom-crafted for Cadillac in 1985 by Hess & Eisenhardt of Cincinnati, Ohio. Beginning with a two-door Coupe DeVille, the car was elongated by adding a set of center door folding jump seats that faced forward, and included a full vinyl roof. These cars were front-wheel drive and the exterior dimensions were smaller than the previous model year. By this time most funeral homes, livery and limousine services were using stretch Fleetwood Brougham rear-wheel drive limousines. These smaller limousines proved to be not as popular as the previous larger cars. In 1985 prices were in the middle to high $30,000.00s.

Here's a three-quarter rear view of the 1985 Fleetwood Factory "75" Sedan and Formal Limousine. Even though it was equipped with all the options Cadillac offered that year, it didn't prove as popular as a Fleetwood Brougham rear-wheel drive stretch. Production figures on this limousine were down dramatically–probably due to the fact that each one had to be custom-crafted by an outside coachbuilder. By 1987 this car had disappeared from the Cadillac line-up and Cadillac turned over all its limousine orders to a wide variety of specialized coachbuilders who could build "full size" limousines using the Fleetwood Brougham.

The 1986 Fleetwood front-wheel drive factory "75" Limousine was almost an identical copy to the 1985 model. The only changes were in the trim and the rear taillights. These limousines didn't prove to be as popular in limousine and livery companies as the stretch coachbuilder-built rear-wheel drive Fleetwood Broughams of the same year. Prices by 1986 were in the high $30,000.00s to low $40,000.00s.

The final year for a factory-built Cadillac Limousine was 1987. These interesting front-wheel drive limousines were actually built for Cadillac by Hess & Eisenhardt of Cincinnati, Ohio. Very few were produced in this last year and Cadillac decided to drop the model and its prestigious nametag after almost a century of use. The only change from previous years was the "crosshatch" grill design in the front.

The 1987 Cadillac Presidential Limousine was actually a concept car designed by Cadillac and built by O'Gara-Hess & Eisenhardt in Cincinnati, Ohio. At least two were produced and traveled the world over, being shown at all international auto shows as well as at home in the U.S.A. Designed to be used by high-powered corporate executives or heads of state, to the author's knowledge, only two were ever produced. Manufacturing cost was prohibitive, and therefore, the project disappeared.

DaBryan Coachbuilder of Springfield, Missouri, manufactured this double-cut 1988 Cadillac Fleetwood Brougham Presidential Limousine photographed in the downtown square in St. Louis, Missouri. Notice that the rear quarter panel has been extended eight inches, giving rear-seat passengers more room and ease of entry and egress. Limousines by DaBryan Coachbuilders are truly built to custom order. Cadillac and Lincoln are their mainstay, but they will also do Rolls-Royce, Mercedes-Benz and Suburbans. "The sky's the limit" here.

1989 would be the last year for this front-end styling for Cadillac. Shown here in front of Limousine Werks showroom in Schaumberg, Illinois, is the 1989 Limousine Werks Corporate Formal Limousine with forward-facing jump seats. These types of limousines were popular with CEOs, bankers, and U.S. government officials. These were considered low-profile or formal limousines. For a much-added expense these cars were often armored discreetly.

This 1990 Miller-Meteor Paramount Cadillac Formal Limousine was hand-crafted by Ron Collins, Collin's Industries in Hutchson, Kansas. The base unit used to create this one-of-a-kind car was the large rear-wheel drive Fleetwood Brougham. How many were made is unknown. The price was probably in the low to middle $40,000.00s.

Parked here is a 1992 Cadillac 26-inch stretch CEO Fleetwood Brougham Limousine built by Classic Coaches of Fountain Valley, California. This limousine was built as a personal car for a discerning individual rather than as a livery vehicle. Production by Classic Coaches was strictly limited. Price was undisclosed.

Chicago Armor and Limousine manufactured to special order this Corporate 34-inch stretch CEO formal limousine. This is a double-cut car (in both the center and the rear quarter panel). Notice the extra side quarter window added to the base Fleetwood Brougham. This car had an electric partition window divider and forward-facing jump seats, which could carry two extra passengers. Cost was in the high $50,000.00s.

S&S/Superior of Ohio started building six-door funeral limousines in the early 1980s. The tradition of excellence continues today. This Sayers & Scoville six-door nine-passenger Cadillac Fleetwood Brougham Family Limousine was parked here for a photo session in 1994. Notice the S&S badging on the rear sail panel of the roof and the wire wheels.

A 24-hour Car is a limousine that can be used for funeral duties as well as VIP use just by flipping the center bench seat the opposite way with a flick of a lever. Here is Superior Coaches 1996 Cadillac Fleetwood Brougham 24-hour Car, actually a "Combination Limousine" used for two or more duties. The center doors were solenoid-operated, giving the appearance of a four-door VIP model.

A pair of 1998 Cadillac DeVille VIP bar cars are seen here in front of a hotel in Ohio. These are 70-inch stretches made by S&S/Superior Coaches of Lima, Ohio. Front-wheel drive with the Northstar engine, these cars boasted ultra luxurious appointments from the coachbuilder. Notice that opera lights are affixed to the center glass panels. These two cars are Superior Coaches' finest "evening out on the town" limousines.

Viking Coachworks of Sanford, Florida, built this short-wheelbase Sedan DeVille front-wheel drive Cadillac Limousine in 1999. The 30-inch stretch housed two companion seats with a center console. This car probably was for an executive or for personal use. Cost was in the middle $50,000.00s.

The newly restyled 2000 Cadillac DeVille with the 32-valve Northstar engine, seen here, is now a six-door family limousine converted by Federal Coach of Fort Smith, Arkansas. This vehicle has three rows of bench-type seating and can carry nine adults in comfort. Primary use is as a funeral family limousine. This car has all the 2000 standard Cadillac features plus an additional row of seats. Cost is in the middle to high $50,000.00s depending on equipment.

The Lincoln Continental Executive Limousine for 1968, by Chicago coachbuilder, Lehmann-Peterson, who started the stretch limousine revolution by building the very first stretch limousines, in 1963, exclusively using Lincolns. By 1968 prices were a little over $15,000.00 and annual production peaked at over 500 units a year. Each was custom-crafted to the end users specifications. The added center section was 36 inches in length and, flanked by two rear-facing companion seats, housed a cocktail cabinet, color TV and other audio/visual controls.

In 1970, a new Lincoln Continental Limousine was offered on the limousine market. It was the A.H.A. Extended Formal. A.H.A. stood for Andy Hotton Associates, located in Belleville, Michigan. Extra length was added to a base Lincoln Continental in the rear doors and quarter panel. Shown here is a 1970 model. This was either available as an eight-passenger sedan without glass divider or a formal limousine with divider. The two jump seats could be made to face forward or rearward.

In 1971 the A.H.A. Lincoln Continental Formal remained virtually unchanged from the previous year. Notice the built-in "Continental Decklid" motif and the reduced privacy rear window. This formal limousine appeared more as a large sedan. This car was equipped with a formal limousine divider window.

Andy Hotton Associates built luxury stretch Lincoln Continental limousines throughout the 1970s and well into the 1980s. Shown here is a 1972 Lincoln Limousine model. For the first three or four years they were almost identical in styling except for the annual grill changes. The company eventually disappeared and then in the 1990s re-emerged under new ownership with center cut stretch styling as most every coachbuilder has today. They are now located in Canada.

This 1977 Lincoln Town Car Executive Limousine is a presidential model built by Executive CoachBuilders of Springfield, Missouri. This car was a 36-inch center cut stretch and housed two rear-facing seats and a center console. Powered by a 460-cubic-inch engine with dual exhaust, it was not exactly economical to operate.

Armbruster-Stageway Inc., located in Fort Smith, Arkansas, has been stretching automobiles since the 1920s. Their specialties are primarily airport cars, big band buses and limousines for the wealthy. Shown is a Town Car six-door limousine built for the funeral industry. Entry and egress were much easier for families in a six-door than a conventional four-door with jump seats. Six-door cars always had three rows of bench seats facing forward.

Shown here in the snow is the 1985 Lincoln Town Car Royale CEO Limousine, manufactured by Royale Coachworks of Haverhill, Massachusetts. This car was produced in very small numbers for whoever wanted a low profile personal limousine. This car was equipped with two small rear-facing folding jump seats for two additional passengers. The stretch length was 26 inches. Price is unknown but it was probably in the high $40,000.00s to middle $50,000.00s.

1986 was an attractive styling year for the Lincoln Town Car as shown here in front of the O'Keefe Centre in Toronto, Canada. This is the very stylish A.H.A. Lincoln Town Car Presidential Limousine. This limousine was the pinnacle of Canadian Coachbuilding at its finest. Since then, there has never been any other Canadian Coachbuilder with such stunning style. Others have tried but failed, due to lack of the style and class that A.H.A. has–they're still in existence today. The National Ballet of Canada was on the playbill at the O'Keefe Centre that evening.

"A very unusual limousine" would definitely be the way to describe this 1986 Lincoln Town Car convertible parade car by KBCC Coachbuilder of Minnesota. It appears that there is no top, just a hard boot around the edge. I hope it doesn't get caught in the rain!

With a 36-inch center stretch, this 1986 Lincoln Mark VI Limousine by Bradford Coachworks of Boco Raton, Florida, proved to be a successful limousine designed for prestigious transportation. The interior featured a rear-facing console with two jump seats facing to the rear providing conference room styling. Not many Mark VI Lincoln Limousines were made compared to the Lincoln Town Car version of the same era.

A 1992 Mini-Lincoln Town Car Limousine made by Classic Coach of Fountain Valley, California, is shown here. Classic Coach does body extensions from 6 to 120 inches and anywhere in between. This CEO corporate limousine undoubtedly went to an executive whose time was too valuable to spend driving himself. Equipped with audio/visual equipment, beverage service, and fax machine, the executive could be productive while being chauffeured to work. The Wall Street Journal was not included in the standard list of options. Notice the television antenna on the trunk and the chrome band on the roof.

A 1994 Lincoln Town Car Formal Limousine from Royale Coachworks of Haverhill, Massachusetts. Royale Coachworks built some of the most interesting "CEO" or formal corporate cars in the 1990s. They were equipped with forward-facing jump seats, no partition and were primarily used for personal executive travel or funeral home use for bereaved families. How many were built is unknown. Here is a 1994 example of a Lincoln interior.

S&S/Superior of Lima, Ohio, also manufactures a line of Lincoln Town Car Limousines and funeral vehicles. Shown here is a 1994 Companion Coachmate Lincoln six-door family limousine built by Sayers & Scoville. Many funeral homes and livery services widely use these types of vehicles today. Here it sits for a photo session just after it came off the line in Lima.

A pair of 1997 Lincoln Town Car Limousines are shown here. One is a six-door used primarily for funeral work. The larger one in the background is a 24-hour Car used for funeral and VIP work. Both were manufactured by Superior of Ohio. The 24-hour Lincoln has solenoid-operated center doors, giving it the appearance of a formal four-door, but with the practicality of six doors for ease of entry and egress.

This Lincoln Town Car Limousine by DaBryan Coachworks of Springfield, Missouri, is a 1997 model. Notice the short version and the long version. Size really does matter as well as personal taste. The short wheelbase model would be considered a CEO or formal personal car while the long wheelbase model would be for a livery or limousine service. Prices varied accordingly!

The 1998 Lincoln VIP Town Car Limousine was a heavy-duty commercial variation of the Lincoln Town Car Sedan. Prices varied from coachbuilder to coachbuilder depending on options and features ordered. In 1998 there were 21 coachbuilders that were Ford certified QVM (qualified vehicle modifier) builders, meaning the Limousine conversion would be warranted by Ford Motor Company. Other builders, even though not certified, could still build limousines, but were not warranted by Ford.

A 1999 Lincoln Town Car six-door funeral family limousine and matching hearse wait outside of a church and cemetery as services conclude. Notice how the coachbuilder's art can make both vehicles compatible as a matched fleet. The limousine's price would be in the low $50,000.00s while the matching Lincoln Hearse would be in the low to middle $60,000.00s. Many funeral homes in the 1990s turned to rear-wheel drive Lincoln Motor equipment.

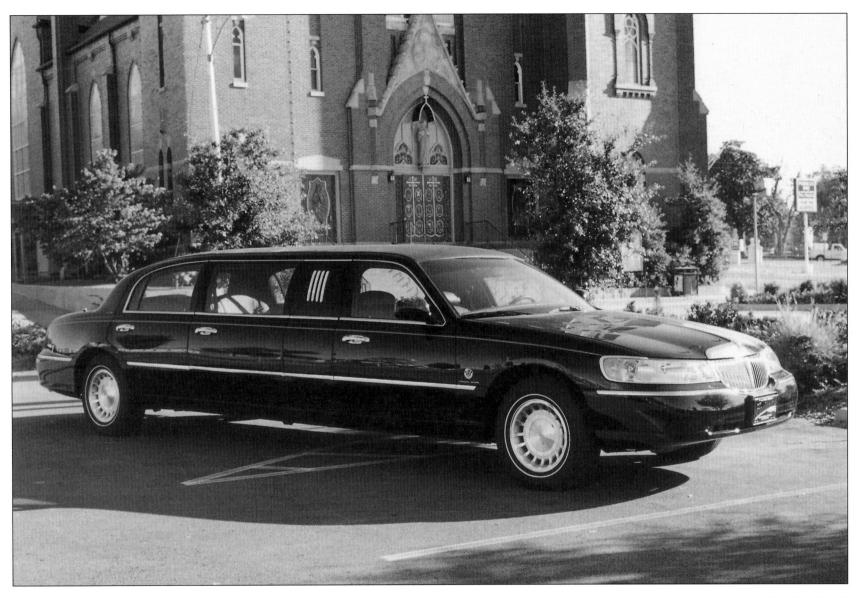

Here is a three-quarter frontal view of the new 2000 Lincoln Town Car six-door 24-hour VIP Limousine by Federal Coach of Fort Smith, Arkansas. This vehicle can be used for funeral work or VIP use, simply by reversing the center flip seat. Federal Coach manufactures a fine line of funeral and VIP limousines and hearses.

Stretched 2000 Lincoln Navigator S.U.V. Limousine and dual tandem Lincoln Town Car Limousine by Ultra Coachworks of California proves to be the "shuttle bus" size limousine of today's trends and are both capable of carrying 12 to 15 people at once. It is probably very hard to make turns at corners, especially in small country cemeteries. These are useful as VIP bar cars for a "night out on the town."

This 1958 Chrysler Imperial Ghia Limousine was one of 31 built that year. These cars were custom-crafted in Torino, Italy, by skilled automobile craftsman. The limousine itself was crafted from two-door Imperial Coupes with convertible frames. All body panels were crafted from scratch by hand.

Helena Rubenstein's 1963 Chrylser Imperial Ghia Crown Limousine had a little bit of different work done. The trunk was raised so that her cosmetics sample cases could fit in the trunk. Notice the TV rabbit ears on the roof. This was one of 13 built in 1963 to special order.

Here is another view of Helena Rubenstein's 1963 Chrysler Imperial Ghia Crown Limousine. Notice the wide white walls, two-tone paint, tinted windows and dual rearview fender-mounted mirrors. Undoubtedly this was luxury at its finest in 1963. Prices for the 1963 limousine started at $18,000.00, and went up with extras added.

This 1964 Chrysler Imperial Ghia Crown Formal Landau Limousine was one of ten made that year. This particular car was delivered to the White House in October 1963 for the Kennedy family. It was to replace the 1961 model that the White House was using at the time.

The 1964 Chrysler Crown Imperial Ghia Limousine came in two versions; available as a six-window town version (which was more common) or as a Formal Landau model, shown here. This model started at $18,000.00 and went up from there. The Landau roof option was undoubtedly thousands more in cost, pushing this car in the $20,000.00s price range, plus then some.

This 1965 Chrysler Imperial Ghia Crown Limousine belonged to noted book author and Nobel Prize winner Pearl S. Buck. This was her personal limousine until her death in 1973 at the age of 81. She authored many books and also established the Pearl S. Buck Foundation in 1964 for children. This limousine is now owned by Professional Car Society member Jan Witte of Chester, New York, and is in immaculate condition.

1965 would prove to be the last year the Chrysler Imperial Ghia Crown Limousine was made. The cost of the limousine had risen 54 percent from 1957 to 1965. The owner of Ghia Coachworks had died and Chrysler felt it could have a domestic model made in the U.S. for less. Ten were produced in 1965 for $18,500.00. Tooling for the limousine was shipped to Barreiros in Spain, where ten more were made in 1966. Other than the front and back grills they were pretty much the same car.

Anna Dodge of the Dodge family was an heiress of the Dodge automotive fortune. This 1965 Imperial Ghia Crown Limousine was specially built for her. Sporting a black leather canopy roof, this was considered to be the Ghia Imperial Crown Town Limousine model. Only ten like this one were made in 1965. Interiors were either gray or beige broadcloth, with black leather in the chauffeur's compartment. The carpeting was of sheared mutton fur!

David Sarnoff, head of RCA Corporation, also owned a 1965 Chrysler Imperial Ghia Crown Limousine. Painted in a two-tone color scheme this car was rather unique, sporting a black leather canopy roof and all the luxury equipment offered in 1965. This limousine was approximately $9,500.00 more than the Cadillac Fleetwood "75" of the same year. This was the last year of Ghia Crowns.

Armbruster-Stageway Coaches of Fort Smith, Arkansas, built this interesting 1967 LeBaron Limousine on a Chrysler Imperial chassis. After the Imperial Ghia Limousine was discontinued in 1965 tooling for these cars was shipped to Barreiros in Spain where ten more cars were built for the European market only in 1966. This was the American answer for a Chrysler Division Limousine in 1967; less than 10 were built and prices were between $12,000.00 and $15,000.00, depending on equipment ordered.

Parked here at curbside is a 1969 Chrysler Imperial LeBaron Limousine by Armbruster-Stageway Coaches of Fort Smith, Arkansas. With only less than 10 of these cars made in 1969 they were dubbed the longest production limousine in the world. Cost was believed to be $16,000.00 plus, depending on options.

Above is one of only two 1972 Hess & Eisenhardt Imperial Chrysler limousines built for the Secret Service fleet. These cars were rolling fortresses on wheels. They were to be used by visiting dignitaries and Heads of State. Truly custom built, these cars were Level VI armored and heavily bullet and bomb proof.

The last Chrysler Imperial Limousine made was by Andy Hotton, Inc. of Belleville, Michigan, in 1982, available as either an executive four-door sedan or a limousine with glass divider window. These cars were hand-crafted from two-door Imperial coupes. It's believed that only two were made and only one exists today on the West Coast in a movie studio.

Phillips Motor Car Corp. of Florida built Chrysler's prototype limousine in 1982-1983 on the LeBaron K-car platform. Exactly how many were built is not known, but they had all the limousine luxuries, only on a much smaller compact scale. The jump seats faced rearward when used and a small console was housed in between them.

The actual LeBaron Limousine for the Chrysler Corporation was produced in small numbers in the middle and late 1980s. Shown here is a 1986 model. These were available as an executive sedan or as a limousine with sliding glass partition divider.

Royale Coach of California built several of these Chrysler LHS New Yorker Limousines for hotels and casinos in Las Vegas. This particular limousine is green in color, has a full padded vinyl roof and is full of luxury interior appointments. Notice the boomerang TV antenna on the trunk lid. This is a 1996 model LHS New Yorker.

Specialty Limousines 1962-1993

This special Buick Electra 225 Limousine built by the Flxible Company of Loudonville, Ohio, contained a pair of forward-facing jump seats, black vinyl roof, and powered center divider window.

In 1962, the Flxible Company of Loudonville, Ohio, built two of these special eight-passenger limousines on the Buick Electra 225 chassis. They were especial ordered limousines; Flxible mainly built ambulances and hearses using the Buick Electra 225 chassis. Only two exist today–one in New York and the other in California. They were in the $10,000.00 price range.

The 1962 Pontiac Limousine was a whole new body style over the previous year. Seating nine adults in comfort, it matched many funeral home fleets who operated Pontiac Superior Funeral Coaches. The cost was around $6,500.00.

Chevrolet Division of General Motors normally did not build a limousine, but National Coaches Inc. of Knightstown, Indiana, would build a limousine, funeral coach or ambulance on any chassis the customer ordered. Shown here is a 1964 Chevrolet Impala Limousine. This firm did conversions mostly for the funeral trade, some airporter stretches and a few custom station wagons in the late 1960s.

National Coaches Inc. of Knightstown, Indiana, was famous for fine handmade cars since 1900. In the 1950s and 1960s they would build to suit any customer's wishes on any chassis desired. Here is a fine example of a 1964 Buick Electra 225 Limousine probably destined for an Indiana funeral home. National's prices for fine craftsmanship were far below that of Cadillac custom coachbuilders.

In its second to the last year of production the luxurious Oldsmobile "98" Cotner-Bevington for 1965 was truly a custom order vehicle for funeral homes that wanted to compliment their Olds hearse. This was a nine-passenger vehicle made in Blytheville, Arkansas. Cotner-Bevington at the time was a part of the Wayne Corporation, a large company that manufactured school buses and ambulances. This car rode on a huge 150-inch wheelbase and retailed for $9,283.00.

Built as a possible production vehicle, this 1965 Ford Galaxie 500 LTD Limousine was specially built by Andrew Hotton Associates of Belleville, Michigan. A show car, it boasted hardtop styling with a full vinyl roof and folding jump seats. Price was close to $3,500.00 in this era.

The 1966 Ford Galaxie 500 LTD Limousine was finally in demand with governmental bodies that wanted to have economical prestige cars that would stay within taxpayer set bounds. With hardtop styling, it provided an interesting alternative to the traditional Cadillac of the same year. Andrew Hotton Associates of Michigan did quite a nice conversion of these cars.

Since the 1940s, Checker Motors Corporation of Kalamazoo, Michigan, produced mostly taxi cabs. True, Checker did build some stretch "airporter" limousines in either a six or eight-door configuration, but for the luxury minded buyer they built this luxury formal Town Custom Limousine on a 129-inch wheelbase. Equipped either with or without a glass partition it boasted a standard array of extras and full size. This is a 1966 version.

CHECKER
Town Custom Limousine

- Built on 129-inch wheelbase for greater interior room, comfort and luxury
- Full size rectangular auxiliary seats
- Choice of Gray Broadcloth or Aerobus Style upholstery
- Driver's partition optional
- All power and automatic options including Air Conditioning available

Checker Motors Corporation, Kalamazoo, Michigan

This 1966 Superior Coach-built Pontiac Embassy nine-passenger limousine was a very dignified and impressive vehicle. Many funeral homes and hotels used these Pontiacs to match their fleets at a less expensive cost factor than Cadillac's. This car came complete with a fully padded roof of vinyl, forward-facing jump seats and an optional partition divider glass between compartments. This car rode on a 146-inch wheelbase and prices were close to $8,200.00.

National Custom Coaches was famous in crafting handmade cars to each customer's specifications and chassis selection. Shown here is a 1966 Buick Electra 225 Custom Formal Landau Limousine. The customer had the option to eliminate the quarter window on the sides of the vehicle and order the Landau roof option for the utmost in privacy.

In 1968, Cotner-Bevington of Arkansas, discontinued Oldsmobile "98" nine-passenger limousines, but still produced Oldsmobile "98" funeral coaches. That did not stop some funeral directors from ordering a matching limousine to compliment their hearses. They came to National Custom Coaches of Knightstown, Indiana, to finish their fleet orders of matching cars. Here is a custom one-off 1968 Oldsmobile "98" nine-passenger limousine photographed outside the factory in Knightstown.

The Stutz Royale Parade Car Limousine for 1970 was indeed quite a special custom creation. The only one produced, its price tag was in excess of $25,000.00 and is believed to be in the Middle East if it still exists today. Notice the forward-facing jump seat used to carry extra passengers in the rear compartment.

The Excalibur Limousine manufactured by the Excalibur Motor Car Corporation of Milwaukee, Wisconsin, retailed in the $150,000.00s. Production numbers were very low and only a few were believed to be produced. Forward or rear-facing jump seats could be ordered to the customer's preferences. This could be considered to be the ultimate American touring limousine of modern times.

In 1980 Bayliff Coachworks of Lima, Ohio, was creating Packard Replicars based on Cadillac running gear and chassis. Bayliff turned out a few of these interesting creations. Shown parked in front of an estate is the Packard Mini-limousine for 1980!

Another Packard Mini-limousine for 1982. Notice how Bayliff Coachbuilders of Lima, Ohio, can take a four-door Cadillac Fleetwood, completely rework it from the ground up, and create a mini-limousine for personal use. Price was believed to be close to $92,000.00 for this true custom car and very few were made. Bayliff mostly produced coupes and convertibles on the new Packard Replicar in the 1980s.

In 1984 luxury included a new Mercedes-Benz 560 SEL Limousine by Moloney Coach of Rolling Meadows, Illinois. Moloney Coachbuilders were direct descendents of Lehmann-Peterson and were famous for Cadillac, Lincoln and government limousines the world over. This Mercedes-Benz 560 SEL Limousine was a 36-inch stretch that included vis-à-vis seating along with a center entertainment console in the center section. Cost was $150.000.00 and production was strictly limited.

Neo-classic cars and limousines became popular in the early 1980s. For timeless styling many limousine operators opted for the Gatsby or neo-classic look. Shown here is an early 1980s DeElegance Limousine by Elegante. Built on a Lincoln Town Car chassis the cost of these cars were usually higher due to the inability to distinguish the year made.

Built for a foreign government, this 1987 Chevrolet Caprice Level VI armored limousine was manufactured by Hess & Eisenhardt of Cincinnati, Ohio. Notice the handrails, custom roof and the formal body style. Flag holders and miniature spot lights on the hood were used for parade occasions. The handrails would be used for bodyguards to hold onto while in a parade.

A 1987 Ford LTD Crown Victoria Limousine custom built by Royale Coachbuilders in California shows an economical way to affordable luxury. Surely this was a "one-off." Royale specialized in one-of-a-kind limousines. The company still exists today.

Built by Armbruster-Stageway of Fort Smith, Arkansas, this airport-style Buick Limousine Station Wagon would be perfect for hotel and shuttle use to airports and back. This late 1980s model had a heavy-duty Buick commercial package frame and suspension to carry the extra load and weight, and so could carry nine passengers and luggage.

The Stutz Royale Head of State Limousine is truly a work of automobile art. Every part, including the rear chrome in the interior, is crafted by hand. Thousands of man-hours went into building this one-of-a-kind limousine. The length is 296 inches. Its wheelbase is 172 inches and engine displacement is 500 inches. This limousine is truly meant for a king! It's believed that at least two exist today in Middle East countries. Price was a whopping $285,000.00 F.O.B. New York, New York.

Limousine Werks of Schaumberg, Illinois, produced a Buick Roadmaster Limousine in 1992 and 1993, until Buick said, "no more" due to lack of certification. Specifications were similar to other expensive models, but thousands of dollars cheaper. This is a four-door VIP model, but a six-door was also available for funeral use. Production numbers were very low.

A 1993 Rolls-Royce Silver Spur touring limousine by Mulliner-Parkward Coachbuilder. These limousines were imported into the USA in 1993, 1994, and 1995. Reportedly prices started at $350,000.00 and went up depending on the options and interior configurations that were ordered. It would take at least six months or longer to obtain this limousine from the coachbuilder, for it was entirely produced by hand. This car took the place of the venerated Phantom VI Rolls-Royce Limousine.

This Cadillac was built by Hess & Eisenhardt in 1956 and photographed by the author in Auburn, Indiana, at the Krause auto auction. This 1956 Cadillac "Queen Mary" Secret Service car was the "chase, or follow-up" car used by the Secret Service from 1956 to 1968 for official White House parades and duties. A chase, or follow-up car follows the President in parades, and carries agents and weapons. This one served several administrations, and was present during the J.F.K. assassination in Dallas, Texas, on November 22, 1963. A private collector now owns the car. This convertible limousine is an important part of current U.S. history.

The 1968 Lincoln Continental Secret Service follow-up car, called the "Queen Mary," was the only 1968 Lincoln convertible made. Regular production of Lincoln convertibles was halted in 1967. This car followed the Presidential limousine in parades. This car was custom designed by Lehmann-Peterson Coachbuilders in Chicago, Illinois. Note the hand rails, running boards, handrail on trunk and fold down rear bumper, not to mention all the advanced electronic security equipment. Secret Service agents used this special car to protect the Chief Executive.

The 1975 Secret Service "follow-up" or "chase car" was custom-crafted by Hess & Eisenhardt of Cincinnati. Based on a 1975 Cadillac Fleetwood "75" limousine, this car followed the Presidential Limousine, with a lot of highly advanced security features. Secret Service agents rode in the front and rear and also on the outside on the side running boards, with handrails. This car was powered by a huge 500-cubic-inch engine.

The Secret Service "Chase Car" used in the late 1970s to follow the Presidential Limousine was this 1979 Cadillac. Based on a Fleetwood "75" Cadillac limousine, once again it was built in Cincinnati, Ohio, by Hess & Eisenhardt, one of the oldest armor car builders in the U.S. This car was powered by a 425-cubic-inch engine. Note the rear top down in the snow when the photo was taken, and the handrails on the trunk lid and doors.

The 1983 Cadillac Presidential parade limousine and Chevy Suburban follow-up vehicle for Secret Service agents pose in front of the White House for an official file photo. Specifications on these and all governmental vehicles are kept highly confidential. You can bet they are all bullet and bomb-proof.

The Secret Service and U.S. government put into service all new security vehicles in 1983, including this Suburban follow-up "war wagon." Built by Hess & Eisenhardt of Ohio, these were virtual fortresses on wheels containing many defensive devices. Notice the handrails, lights, siren, and tie-down hooks under the front and rear of the vehicle. This was so it could be shipped anywhere in the world via air transport.

The last open top or convertible follow-up car made was this 1984 Cadillac Fleetwood "75" Limousine conversion. Specially built for the Secret Service & U.S. Government, Hess & Eisenhardt produced a lot of these convertible limousine conversions for the U.S. Government and also foreign governments the world over. Here you see one with the rear top up, handrails, flag holders, spotlights on front fenders, and of course running boards. Many interior features were, of course, classified.

The most current Secret Service chase car or follow-up car is this 1989 Cadillac Fleetwood Brougham chassis with raised roof and extended body. This car was a participant in the Clinton Presidential Inauguration parade of 1993. Referred to by Secret Service agents as the "War Wagon," O'Gara-Hess & Eisenhardt manufactured an unknown number of these unusual cars for the U.S. government in the late 1980s and early 1990s.

A Secret Service armor-plated 1987 Cadillac Formal Fleetwood Limousine outside the United Nations in New York City. Notice the thick window frames to hold together the thick multi-layered glass. This limousine was built by Moloney Armouring Coachbuilders of Bensenville, Illinois.

Here is a rear three-quarter view of a U.S. Secret Service Cadillac Fleetwood Formal Limousine featuring double framed side glass and a "privacy" rear window parked outside of the UN in New York City. Built by Moloney Armouring Coachbuilders of Bensenville, Illinois, circa 1988.

Traveling on the streets of New York City, heading to the Seagrams building on Park Avenue, is a Secret Service 1991 Cadillac Fleetwood Formal Armoured Limousine. Notice the 1984 roof style grafted on to a 1991 body style. Crafted by Moloney Armouring of Bensenville, Illinois, who builds exclusively for governments; domestic and foreign.

Old Cadillac Coupe Deville dies may have helped Moloney Coachbuilder of Bensenville, Illinois, build this 1991 34-foot Corporate Formal Cadillac Secret Service Limousine, seen here on the streets of New York City in 1998.

Seen here outside of the Waldorf-Astoria hotel in New York City, in March of 1998, is this Moloney Cadillac Fleetwood Formal Secret Service Limousine for 1992. This hotel attracts many armored vehicles when the UN is in session or the President is in town.

An armor-plated 1992 Cadillac Fleetwood Formal Secret Service Limousine is seen parked here on New York City's 2nd Avenue. Photographed June 25, 1997, this car was being used by the U.S. Ambassador to the UN in New York. Most of these cars are heavily armored, with 2 forward-facing jump seats, rear air, and stereo. Another Moloney creation in protective level's II to VI Armouring.

A 1994 Secret Service Limousine Werks armoured 31-foot stretch Cadillac Fleetwood Brougham Formal Limousine is seen here outside UN Plaza Hotel, in New York City on June 6, 1997. Notice the reduced rear privacy window and small quarter window in the rear door and quarter sail panel.

A side view of a 31-foot stretch Cadillac Fleetwood Formal Limousine by Limousine Werks of Schaumberg, Illinois, shows that Fleetwood Cadillac Formal Limousines can still be made on special order. Most of these cars are armored for the government, however, this 1995 model is unarmored. Notice the side quarter window in the rear door and the small formal rear side quarter window.

An armored 1996 Cadillac Fleetwood Brougham Formal Limousine by Moloney Armouring Coachbuilders of Bensenville, Illinois, is seen here in New York. These cars are extensively used by the Clinton family, the Vice-President's family, UN Ambassadors and anyone else of political importance. The Secret Service maintains a fleet of these limousines all over the U.S. and the world.

An armored 1996 Cadillac Fleetwood Brougham Formal Limousine by Moloney Armouring is being used during Hillary Clinton's visit to Newburgh, New York, for a speech she made there on July 14, 1998. Though heavily armored, one can't tell that this is an armored car. Looks like a Fleetwood "75" style roof on a 26-inch stretch armored limousine.

WASHINGTON — Lincoln Continental perpetuates its title as the "Car of Presidents" with announcement that a new Presidential Limousine (top photo) presently is in service at the White House [1969]. The 1961 limousine (second from top) was used by Presidents Kennedy and Johnson and Heads of State from all over the world. The oversized 1950 Lincoln (second from bottom) served Presidents Truman and Eisenhower. When President Eisenhower took office he suggested that a clear plastic roof be fabricated so that the President could see and be seen in bad weather – thus was born the famous Lincoln "Bubbletop." The "Sunshine Special" (bottom) was a specially built 1939 Lincoln used by President Roosevelt. Lincolns have served the White House since the term of Calvin Coolidge.

The 1961 Lincoln Continental JFK limousine was custom-crafted from a regular Lincoln convertible and made its debut in June of 1961. Hess & Eisenhardt of Cincinnati, Ohio, was the builder. Reportedly over $200,000.00 was invested in making this presidential limousine that was used on that fateful day in Dallas, Texas, in 1963. The limousine had interchangeable roof panels, power operated rear seat, glass partition, advanced communications, and two folding forward-facing jump seats among many other special features. The car is now a part of the collection in the Henry Ford Museum in Dearborn, Michigan.

The 1961 JFK car was transformed into a heavily armored 1964 Lincoln Presidential Limousine for President Johnson, by Hess & Eisenhardt. "D-2 Quick Fix" was its code name while being reworked into a 1964. It is an educated guess that several hundred thousand dollars went into restoring and armoring this limousine. The car weighed close to 10,000 pounds with all the armor and special equipment that was added. This car now has been on display at the Henry Ford Museum since 1977, after it was retired from official White House duty.

The new 1969 Lincoln Continental Executive Presidential Limousine for the White House was delivered to the Secret Service in October of 1968. Custom-built by Chicago coachbuilders Lehmann-Peterson, this car was based on a 1967 Lincoln Platform, but for two years had updated exterior trim added until it was delivered in October of 1968. The car was 21 feet long, weighed 6 tons, and rode on a 160-inch wheelbase. It was reportedly a half million dollars in cost to build. Two tons of armor kept the occupants definitely bullet and bomb-proof, and safe.

This specially constructed 1969 Lincoln Continental Executive Limousine had more advanced security, communications and engineering features than any other automobile ever used for official duties at the White House. When the President was riding in the car, the flag was flown from the left fender. A center section of the glass roof was hinged so that it could be opened, permitting occupants to stand in the passenger compartment during parades. The rear bumper could be lowered like a tailgate and converted into a platform for Secret Service agents. An adjustable handrail was raised hydraulically from its recessed location in the trunk deck. The new vehicle supplements the 1961 Lincoln Continental Presidential Limousine that was in service at the White House.

This 22-foot custom-built black Lincoln Continental, designed and built by Ford Motor Company, was the latest in a long line of White House Lincolns which have earned the title of "Car of Presidents." The limousine had a roof section which opened to permit two occupants to stand in the passenger compartment during parades. The car had a wheelbase of 161 inches and was powered by a 460-cubic-inch Lincoln engine. The roof of the new limousine was approximately 2.5 inches higher than the regular production Lincoln Continental and the body was 34 inches longer. It served Presidents Nixon and Ford during their administration.

Added to the Presidential fleet of limousines in 1972 was this 22-foot custom-built black Lincoln Continental Limousine powered by a 460-cubic-inch Lincoln engine with a wheelbase of 161 inches. This was truly a car "fit for a President." President Nixon and President Ford primarily used this vehicle for official parades and White House duties. Note the Lincoln suicide doors used, instead of standard 1972 Lincoln rear doors. Equipped with all the latest security advances, it reportedly cost over $500,000 to build.

This 1983 Cadillac Presidential Limousine, which was highlighted by a black exterior paint finish, was a specially built version of Cadillac's production limousine. It featured a special chassis and powertrain designed and built by Cadillac in cooperation with the GM Engineering Staff. Special armoring and body modifications were designed and developed by Hess & Eisenhardt Armouring of Cincinnati–a division of O'Gara International Ltd. This Presidential limousine had a wheelbase of 161.5 inches, 17 inches longer than the conventional Cadillac limousine, which allowed seating for five persons in the rear compartment when the folding seats were occupied. The interior was appointed with dark blue Prima cloth trim. A large displacement V-8 engine, built by Cadillac, powered this limousine.

Another new Presidential Limousine joined the White House fleet of cars in January of 1984. This specially built 1983 Cadillac Presidential edition Fleetwood Limousine was the latest edition to the White House protective limousine fleet. Built by Hess & Eisenhardt Armouring of Cincinnati, it incorporated all the latest armoring and security features available. Reportedly two were produced, so that one could be "leap-frogged" ahead to another major city or town on Airforce One for the President when he arrived. President Reagan was the Limousine's primary user, and its delivery to the White House coincided with his birthday in 1984.

Lincoln perpetuates its title as the "Car of Presidents" with the announcement that a new Presidential limousine has been delivered to the U.S. Secret Service. This 1989 Lincoln Town Car was custom built in Dearborn Heights, Michigan. The one-of-a-kind limousine maintained the interior and exterior styling theme of the production Lincoln Town Car, yet incorporated into its advanced design security, communications and engineering features.

In early January, 1993, the first of three identical "Cadillac Fleetwood Brougham-Presidential Series" limousines was delivered to the White House; just in time for President Bill Clinton's Inauguration parade. This Limousine rode on a total length of 270 inches with a wheelbase of 167.5 inches, and weighed 6 tons. The limousine for Bill Clinton was built totally "in-house" by Cadillac Motor Car Division and GM's advanced engineering staff. The next two cars were delivered shortly afterward in the next few months of 1993. They are black with blue cloth interior with ZeBrano wood accents. They are powered by GM's 5.7-liter 350-cubic-inch engine. Many advanced security items and electronic equipment are on board, not to mention twelve interior speakers for premium sound. The reported cost for all three Limousines is close to seven million dollars. An unconfirmed rumor has it that a 2001 Cadillac Presidential Limousine is being built by January 2001, for the next President of the U.S.

Meet the Author

The author–a Professional Car Society member–with his 1979 Cadillac Miller-Meteor Funeral Coach.

Richard J. Conjalka has had a fascination for motorized professional funeral and limousine vehicles since 1959, when he was eight years of age.

This unique interest began when young Richard was handed a large stack of funeral home trade magazines from a local funeral director who was cleaning house. After clipping all the automotive ads from the magazines, he was hooked on the subject. From that point on, he became a firm collector of material on these interesting but "not talked about" cars.

As of this date, he is credited with having one of the largest collections in the Midwestern United States, of hearse, ambulance and limousine material, photographs, literature, and advertising to exist. The contents of this book though covering the major evolution of the limousine, barely scratches the surface of Mr. Conjalka's collection.

Retired after working in the funeral industry for many years, he now spends much of his time at the golf course and traveling. He resides in northwest Indiana. Among his interests are professional cars, limousines, photography, golf and history, all of which will hold him in good stead as he continues to work on other projects in the future.

MORE TITLES FROM ICONOGRAFIX:

All Iconografix books are available from direct mail specialty book dealers and bookstores worldwide, or can be ordered from the publisher. For book trade and distribution information or to add your name to our mailing list and receive a **FREE CATALOG** contact:

Iconografix, PO Box 446, Hudson, Wisconsin, 54016   Telephone: (715) 381-9755, (800) 289-3504 (USA), Fax: (715) 381-9756

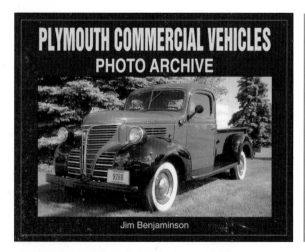
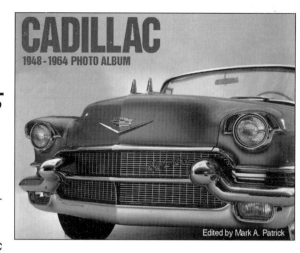

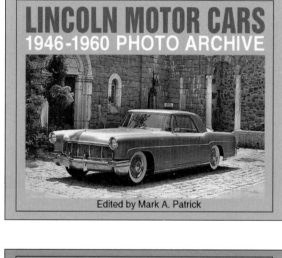
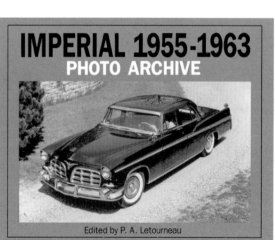
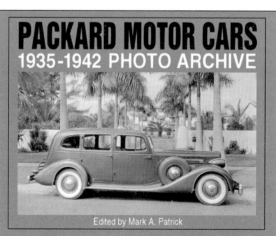